DRONE PILOT LOG

Pilot Name   :

Address   :

Phone No.   :

E-mail   :

Certificates :   Ratings   :
              :   :
              :   :

All rights reserved. No portion of this book may be reproduced, stored in a retrieval system, or transmitted in any form or by any means—electronic, mechanical, photocopy, recording, scanning, or other, without the prior written permission of the publisher and/or authors.

ISBN-13: 978-1503000322
ISBN-10: 150300032X

Copyright © 2014 Drone Academy X
All rights reserved.

DRONE PILOT LOG

DRONE X ACADEMY

CONTENTS	PAGE
DRONE PILOT LOG SHEETS	4
DRONE OPERATION DRAWING SHEETS	52

DRONE PILOT LOG

DATE	FLIGHT NO.	DRONE INFORMATION		FLIGHT LOCATION		FLIGHT CONDTIONS
		DRONE MAKE AND MODEL	DRONE ID NO.	TAKEOFF	LANDING	WEATHER, WIND, TEMPERATURE, ETC.

COMMENTS

REFERENCE KEY
CR — Calibration Required
HZ — Hazard
PP — People and Property
AT — Air Traffic
MI — Mechanical Issue
TI — Technical Issue
PIC — Pilot in Command

OPERATIONAL ISSUES						REMARKS & COMMENTS	PILOTING TIME		
CR	HZ	PP	AT	MI	TI		IN	OUT	TOTAL
						TOTAL HOURS THIS PAGE			
						TOTAL FORWARD			
						TOTAL TO DATE			

I CERTIFY THAT THE FOREGOING ENTRIES ARE TRUE AND CORRECT:

BY: _____   DATE: _____

DRONE PILOT LOG

DATE	FLIGHT NO.	DRONE INFORMATION		FLIGHT LOCATION		FLIGHT CONDTIONS
		DRONE MAKE AND MODEL	DRONE ID NO.	TAKEOFF	LANDING	WEATHER, WIND, TEMPERATURE, ETC.

COMMENTS

REFERENCE KEY
CR   Calibration Required
HZ   Hazard
PP   People and Property
AT   Air Traffic
MI   Mechanical Issue
TI   Technical Issue
PIC   Pilot in Command

OPERATIONAL ISSUES						REMARKS & COMMENTS	PILOTING TIME		
CR	HZ	PP	AT	MI	TI		IN	OUT	TOTAL
						TOTAL HOURS THIS PAGE			
						TOTAL FORWARD			
						TOTAL TO DATE			

I CERTIFY THAT THE FOREGOING ENTRIES ARE TRUE AND CORRECT:

BY: _____   DATE: _____

DRONE PILOT LOG

DATE	FLIGHT NO.	DRONE INFORMATION		FLIGHT LOCATION		FLIGHT CONDTIONS
		DRONE MAKE AND MODEL	DRONE ID NO.	TAKEOFF	LANDING	WEATHER, WIND, TEMPERATURE, ETC.

COMMENTS

REFERENCE KEY
CR   Calibration Required
HZ   Hazard
PP   People and Property
AT   Air Traffic
MI   Mechanical Issue
TI   Technical Issue
PIC   Pilot in Command

OPERATIONAL ISSUES						REMARKS & COMMENTS	PILOTING TIME		
CR	HZ	PP	AT	MI	TI		IN	OUT	TOTAL
						TOTAL HOURS THIS PAGE			
						TOTAL FORWARD			
						TOTAL TO DATE			

I CERTIFY THAT THE FOREGOING ENTRIES ARE TRUE AND CORRECT:

BY: _____   DATE: _____

DRONE PILOT LOG

DATE	FLIGHT NO.	DRONE INFORMATION		FLIGHT LOCATION		FLIGHT CONDTIONS
		DRONE MAKE AND MODEL	DRONE ID NO.	TAKEOFF	LANDING	WEATHER, WIND, TEMPERATURE, ETC.

COMMENTS

REFERENCE KEY
CR   Calibration Required
HZ   Hazard
PP   People and Property
AT   Air Traffic
MI   Mechanical Issue
TI   Technical Issue
PIC   Pilot in Command

OPERATIONAL ISSUES						REMARKS & COMMENTS	PILOTING TIME		
CR	HZ	PP	AT	MI	TI		IN	OUT	TOTAL
						TOTAL HOURS THIS PAGE			
						TOTAL FORWARD			
						TOTAL TO DATE			

I CERTIFY THAT THE FOREGOING ENTRIES ARE TRUE AND CORRECT:

BY: _____   DATE: _____

DRONE PILOT LOG

DATE	FLIGHT NO.	DRONE INFORMATION		FLIGHT LOCATION		FLIGHT CONDTIONS
		DRONE MAKE AND MODEL	DRONE ID NO.	TAKEOFF	LANDING	WEATHER, WIND, TEMPERATURE, ETC.

COMMENTS

REFERENCE KEY
CR   Calibration Required
HZ   Hazard
PP   People and Property
AT   Air Traffic
MI   Mechanical Issue
TI   Technical Issue
PIC   Pilot in Command

| OPERATIONAL ISSUES |||||| REMARKS & COMMENTS | PILOTING TIME |||
CR	HZ	PP	AT	MI	TI		IN	OUT	TOTAL
						TOTAL HOURS THIS PAGE			
						TOTAL FORWARD			
						TOTAL TO DATE			

I CERTIFY THAT THE FOREGOING ENTRIES ARE TRUE AND CORRECT:

BY: _____   DATE: _____

DRONE PILOT LOG

DATE	FLIGHT NO.	DRONE INFORMATION		FLIGHT LOCATION		FLIGHT CONDTIONS
		DRONE MAKE AND MODEL	DRONE ID NO.	TAKEOFF	LANDING	WEATHER, WIND, TEMPERATURE, ETC.

COMMENTS

REFERENCE KEY
CR   Calibration Required
HZ   Hazard
PP   People and Property
AT   Air Traffic
MI   Mechanical Issue
TI   Technical Issue
PIC   Pilot in Command

OPERATIONAL ISSUES						REMARKS & COMMENTS	PILOTING TIME		
CR	HZ	PP	AT	MI	TI		IN	OUT	TOTAL
						TOTAL HOURS THIS PAGE			
						TOTAL FORWARD			
						TOTAL TO DATE			

I CERTIFY THAT THE FOREGOING ENTRIES ARE TRUE AND CORRECT:

BY: _____ DATE: _____

DRONE PILOT LOG

DATE	FLIGHT NO.	DRONE INFORMATION		FLIGHT LOCATION		FLIGHT CONDTIONS
		DRONE MAKE AND MODEL	DRONE ID NO.	TAKEOFF	LANDING	WEATHER, WIND, TEMPERATURE, ETC.

COMMENTS

REFERENCE KEY
CR   Calibration Required
HZ   Hazard
PP   People and Property
AT   Air Traffic
MI   Mechanical Issue
TI   Technical Issue
PIC   Pilot in Command

OPERATIONAL ISSUES						REMARKS & COMMENTS	PILOTING TIME		
CR	HZ	PP	AT	MI	TI		IN	OUT	TOTAL
						TOTAL HOURS THIS PAGE			
						TOTAL FORWARD			
						TOTAL TO DATE			

I CERTIFY THAT THE FOREGOING ENTRIES ARE TRUE AND CORRECT:

BY: _____ DATE: _____

DRONE PILOT LOG

DATE	FLIGHT NO.	DRONE INFORMATION		FLIGHT LOCATION		FLIGHT CONDTIONS
		DRONE MAKE AND MODEL	DRONE ID NO.	TAKEOFF	LANDING	WEATHER, WIND, TEMPERATURE, ETC.

COMMENTS

REFERENCE KEY
CR   Calibration Required
HZ   Hazard
PP   People and Property
AT   Air Traffic
MI   Mechanical Issue
TI   Technical Issue
PIC   Pilot in Command

OPERATIONAL ISSUES						REMARKS & COMMENTS	PILOTING TIME		
CR	HZ	PP	AT	MI	TI		IN	OUT	TOTAL
						TOTAL HOURS THIS PAGE			
						TOTAL FORWARD			
						TOTAL TO DATE			

I CERTIFY THAT THE FOREGOING ENTRIES ARE TRUE AND CORRECT:

BY: _____   DATE: _____

DRONE PILOT LOG

DATE	FLIGHT NO.	DRONE INFORMATION		FLIGHT LOCATION		FLIGHT CONDTIONS
		DRONE MAKE AND MODEL	DRONE ID NO.	TAKEOFF	LANDING	WEATHER, WIND, TEMPERATURE, ETC.

COMMENTS

REFERENCE KEY
CR Calibration Required
HZ Hazard
PP People and Property
AT Air Traffic
MI Mechanical Issue
TI Technical Issue
PIC Pilot in Command

OPERATIONAL ISSUES						REMARKS & COMMENTS	PILOTING TIME		
CR	HZ	PP	AT	MI	TI		IN	OUT	TOTAL
						TOTAL HOURS THIS PAGE			
						TOTAL FORWARD			
						TOTAL TO DATE			

I CERTIFY THAT THE FOREGOING ENTRIES ARE TRUE AND CORRECT:

BY: _____ DATE: _____

DRONE PILOT LOG

DATE	FLIGHT NO.	DRONE INFORMATION		FLIGHT LOCATION		FLIGHT CONDTIONS
		DRONE MAKE AND MODEL	DRONE ID NO.	TAKEOFF	LANDING	WEATHER, WIND, TEMPERATURE, ETC.

COMMENTS

REFERENCE KEY
CR   Calibration Required
HZ   Hazard
PP   People and Property
AT   Air Traffic
MI   Mechanical Issue
TI   Technical Issue
PIC   Pilot in Command

OPERATIONAL ISSUES						REMARKS & COMMENTS	PILOTING TIME		
CR	HZ	PP	AT	MI	TI		IN	OUT	TOTAL
						TOTAL HOURS THIS PAGE			
						TOTAL FORWARD			
						TOTAL TO DATE			

I CERTIFY THAT THE FOREGOING ENTRIES ARE TRUE AND CORRECT:

BY: _____ DATE: _____

DRONE PILOT LOG

DATE	FLIGHT NO.	DRONE INFORMATION		FLIGHT LOCATION		FLIGHT CONDTIONS
		DRONE MAKE AND MODEL	DRONE ID NO.	TAKEOFF	LANDING	WEATHER, WIND, TEMPERATURE, ETC.

COMMENTS

REFERENCE KEY
CR   Calibration Required
HZ   Hazard
PP   People and Property
AT   Air Traffic
MI   Mechanical Issue
TI   Technical Issue
PIC   Pilot in Command

OPERATIONAL ISSUES						REMARKS & COMMENTS	PILOTING TIME		
CR	HZ	PP	AT	MI	TI		IN	OUT	TOTAL
						TOTAL HOURS THIS PAGE			
						TOTAL FORWARD			
						TOTAL TO DATE			

I CERTIFY THAT THE FOREGOING ENTRIES ARE TRUE AND CORRECT:

BY: _____   DATE: _____

DRONE PILOT LOG

DATE	FLIGHT NO.	DRONE INFORMATION		FLIGHT LOCATION		FLIGHT CONDTIONS
		DRONE MAKE AND MODEL	DRONE ID NO.	TAKEOFF	LANDING	WEATHER, WIND, TEMPERATURE, ETC.

COMMENTS

REFERENCE KEY
CR   Calibration Required
HZ   Hazard
PP   People and Property
AT   Air Traffic
MI   Mechanical Issue
TI   Technical Issue
PIC   Pilot in Command

OPERATIONAL ISSUES						REMARKS & COMMENTS	PILOTING TIME		
CR	HZ	PP	AT	MI	TI		IN	OUT	TOTAL
						TOTAL HOURS THIS PAGE			
						TOTAL FORWARD			
						TOTAL TO DATE			

I CERTIFY THAT THE FOREGOING ENTRIES ARE TRUE AND CORRECT:

BY: _____   DATE: _____

DRONE PILOT LOG

DATE	FLIGHT NO.	DRONE INFORMATION		FLIGHT LOCATION		FLIGHT CONDTIONS
		DRONE MAKE AND MODEL	DRONE ID NO.	TAKEOFF	LANDING	WEATHER, WIND, TEMPERATURE, ETC.

COMMENTS

REFERENCE KEY
CR   Calibration Required
HZ   Hazard
PP   People and Property
AT   Air Traffic
MI   Mechanical Issue
TI   Technical Issue
PIC   Pilot in Command

OPERATIONAL ISSUES						REMARKS & COMMENTS	PILOTING TIME		
CR	HZ	PP	AT	MI	TI		IN	OUT	TOTAL
						TOTAL HOURS THIS PAGE			
						TOTAL FORWARD			
						TOTAL TO DATE			

I CERTIFY THAT THE FOREGOING ENTRIES ARE TRUE AND CORRECT:

BY: _____   DATE: _____

DRONE PILOT LOG

DATE	FLIGHT NO.	DRONE INFORMATION		FLIGHT LOCATION		FLIGHT CONDTIONS
		DRONE MAKE AND MODEL	DRONE ID NO.	TAKEOFF	LANDING	WEATHER, WIND, TEMPERATURE, ETC.

COMMENTS

REFERENCE KEY
CR   Calibration Required
HZ   Hazard
PP   People and Property
AT   Air Traffic
MI   Mechanical Issue
TI   Technical Issue
PIC Pilot in Command

OPERATIONAL ISSUES						REMARKS & COMMENTS	PILOTING TIME		
CR	HZ	PP	AT	MI	TI		IN	OUT	TOTAL
						TOTAL HOURS THIS PAGE			
						TOTAL FORWARD			
						TOTAL TO DATE			

I CERTIFY THAT THE FOREGOING ENTRIES ARE TRUE AND CORRECT:

BY: _____ DATE: _____

DRONE PILOT LOG

DATE	FLIGHT NO.	DRONE INFORMATION		FLIGHT LOCATION		FLIGHT CONDTIONS
		DRONE MAKE AND MODEL	DRONE ID NO.	TAKEOFF	LANDING	WEATHER, WIND, TEMPERATURE, ETC.

COMMENTS

REFERENCE KEY
CR   Calibration Required
HZ   Hazard
PP   People and Property
AT   Air Traffic
MI   Mechanical Issue
TI   Technical Issue
PIC Pilot in Command

OPERATIONAL ISSUES						REMARKS & COMMENTS	PILOTING TIME		
CR	HZ	PP	AT	MI	TI		IN	OUT	TOTAL
						TOTAL HOURS THIS PAGE			
						TOTAL FORWARD			
						TOTAL TO DATE			

I CERTIFY THAT THE FOREGOING ENTRIES ARE TRUE AND CORRECT:

BY: _____   DATE: _____

DRONE PILOT LOG

DATE	FLIGHT NO.	DRONE INFORMATION		FLIGHT LOCATION		FLIGHT CONDTIONS
		DRONE MAKE AND MODEL	DRONE ID NO.	TAKEOFF	LANDING	WEATHER, WIND, TEMPERATURE, ETC.

COMMENTS

REFERENCE KEY
CR   Calibration Required
HZ   Hazard
PP   People and Property
AT   Air Traffic
MI   Mechanical Issue
TI   Technical Issue
PIC   Pilot in Command

OPERATIONAL ISSUES						REMARKS & COMMENTS	PILOTING TIME		
CR	HZ	PP	AT	MI	TI		IN	OUT	TOTAL
						TOTAL HOURS THIS PAGE			
						TOTAL FORWARD			
						TOTAL TO DATE			

I CERTIFY THAT THE FOREGOING ENTRIES ARE TRUE AND CORRECT:

BY: _____ DATE: _____

DRONE PILOT LOG

DATE	FLIGHT NO.	DRONE INFORMATION		FLIGHT LOCATION		FLIGHT CONDTIONS
		DRONE MAKE AND MODEL	DRONE ID NO.	TAKEOFF	LANDING	WEATHER, WIND, TEMPERATURE, ETC.

COMMENTS

REFERENCE KEY
CR   Calibration Required
HZ   Hazard
PP   People and Property
AT   Air Traffic
MI   Mechanical Issue
TI   Technical Issue
PIC   Pilot in Command

OPERATIONAL ISSUES						REMARKS & COMMENTS	PILOTING TIME		
CR	HZ	PP	AT	MI	TI		IN	OUT	TOTAL
						TOTAL HOURS THIS PAGE			
						TOTAL FORWARD			
						TOTAL TO DATE			

I CERTIFY THAT THE FOREGOING ENTRIES ARE TRUE AND CORRECT:

BY: _____   DATE: _____

DRONE PILOT LOG

DATE	FLIGHT NO.	DRONE INFORMATION		FLIGHT LOCATION		FLIGHT CONDTIONS
		DRONE MAKE AND MODEL	DRONE ID NO.	TAKEOFF	LANDING	WEATHER, WIND, TEMPERATURE, ETC.

COMMENTS

REFERENCE KEY
CR   Calibration Required
HZ   Hazard
PP   People and Property
AT   Air Traffic
MI   Mechanical Issue
TI   Technical Issue
PIC   Pilot in Command

OPERATIONAL ISSUES						REMARKS & COMMENTS	PILOTING TIME		
CR	HZ	PP	AT	MI	TI		IN	OUT	TOTAL
						TOTAL HOURS THIS PAGE			
						TOTAL FORWARD			
						TOTAL TO DATE			

I CERTIFY THAT THE FOREGOING ENTRIES ARE TRUE AND CORRECT:

BY: _____ DATE: _____

DRONE PILOT LOG

DATE	FLIGHT NO.	DRONE INFORMATION		FLIGHT LOCATION		FLIGHT CONDTIONS
		DRONE MAKE AND MODEL	DRONE ID NO.	TAKEOFF	LANDING	WEATHER, WIND, TEMPERATURE, ETC.

COMMENTS

REFERENCE KEY
CR   Calibration Required
HZ   Hazard
PP   People and Property
AT   Air Traffic
MI   Mechanical Issue
TI   Technical Issue
PIC   Pilot in Command

OPERATIONAL ISSUES						REMARKS & COMMENTS	PILOTING TIME		
CR	HZ	PP	AT	MI	TI		IN	OUT	TOTAL
						TOTAL HOURS THIS PAGE			
						TOTAL FORWARD			
						TOTAL TO DATE			

I CERTIFY THAT THE FOREGOING ENTRIES ARE TRUE AND CORRECT:

BY: _____   DATE: _____

DRONE PILOT LOG

DATE	FLIGHT NO.	DRONE INFORMATION		FLIGHT LOCATION		FLIGHT CONDTIONS
		DRONE MAKE AND MODEL	DRONE ID NO.	TAKEOFF	LANDING	WEATHER, WIND, TEMPERATURE, ETC.

COMMENTS

REFERENCE KEY
CR   Calibration Required
HZ   Hazard
PP   People and Property
AT   Air Traffic
MI   Mechanical Issue
TI   Technical Issue
PIC   Pilot in Command

OPERATIONAL ISSUES						REMARKS & COMMENTS	PILOTING TIME		
CR	HZ	PP	AT	MI	TI		IN	OUT	TOTAL
						TOTAL HOURS THIS PAGE			
						TOTAL FORWARD			
						TOTAL TO DATE			

I CERTIFY THAT THE FOREGOING ENTRIES ARE TRUE AND CORRECT:

BY: _____ DATE: _____

DRONE PILOT LOG

DATE	FLIGHT NO.	DRONE INFORMATION		FLIGHT LOCATION		FLIGHT CONDTIONS
		DRONE MAKE AND MODEL	DRONE ID NO.	TAKEOFF	LANDING	WEATHER, WIND, TEMPERATURE, ETC.

COMMENTS

REFERENCE KEY
CR   Calibration Required
HZ   Hazard
PP   People and Property
AT   Air Traffic
MI   Mechanical Issue
TI   Technical Issue
PIC   Pilot in Command

OPERATIONAL ISSUES						REMARKS & COMMENTS	PILOTING TIME		
CR	HZ	PP	AT	MI	TI		IN	OUT	TOTAL
						TOTAL HOURS THIS PAGE			
						TOTAL FORWARD			
						TOTAL TO DATE			

I CERTIFY THAT THE FOREGOING ENTRIES ARE TRUE AND CORRECT:

BY: _____ DATE: _____

DRONE PILOT LOG

DATE	FLIGHT NO.	DRONE INFORMATION		FLIGHT LOCATION		FLIGHT CONDTIONS
		DRONE MAKE AND MODEL	DRONE ID NO.	TAKEOFF	LANDING	WEATHER, WIND, TEMPERATURE, ETC.

COMMENTS

REFERENCE KEY
CR   Calibration Required
HZ   Hazard
PP   People and Property
AT   Air Traffic
MI   Mechanical Issue
TI   Technical Issue
PIC Pilot in Command

OPERATIONAL ISSUES						REMARKS & COMMENTS	PILOTING TIME		
CR	HZ	PP	AT	MI	TI		IN	OUT	TOTAL
						TOTAL HOURS THIS PAGE			
						TOTAL FORWARD			
						TOTAL TO DATE			

I CERTIFY THAT THE FOREGOING ENTRIES ARE TRUE AND CORRECT:

BY: _____   DATE: _____

DRONE PILOT LOG

DATE	FLIGHT NO.	DRONE INFORMATION		FLIGHT LOCATION		FLIGHT CONDTIONS
		DRONE MAKE AND MODEL	DRONE ID NO.	TAKEOFF	LANDING	WEATHER, WIND, TEMPERATURE, ETC.

COMMENTS

REFERENCE KEY
CR Calibration Required
HZ Hazard
PP People and Property
AT Air Traffic
MI Mechanical Issue
TI Technical Issue
PIC Pilot in Command

OPERATIONAL ISSUES						REMARKS & COMMENTS	PILOTING TIME		
CR	HZ	PP	AT	MI	TI		IN	OUT	TOTAL
						TOTAL HOURS THIS PAGE			
						TOTAL FORWARD			
						TOTAL TO DATE			

I CERTIFY THAT THE FOREGOING ENTRIES ARE TRUE AND CORRECT:

BY: _____   DATE: _____

DRONE PILOT LOG

DATE	FLIGHT NO.	DRONE INFORMATION		FLIGHT LOCATION		FLIGHT CONDTIONS
		DRONE MAKE AND MODEL	DRONE ID NO.	TAKEOFF	LANDING	WEATHER, WIND, TEMPERATURE, ETC.

COMMENTS

REFERENCE KEY
CR   Calibration Required
HZ   Hazard
PP   People and Property
AT   Air Traffic
MI   Mechanical Issue
TI   Technical Issue
PIC Pilot in Command

OPERATIONAL ISSUES						REMARKS & COMMENTS	PILOTING TIME		
CR	HZ	PP	AT	MI	TI		IN	OUT	TOTAL
						TOTAL HOURS THIS PAGE			
						TOTAL FORWARD			
						TOTAL TO DATE			

I CERTIFY THAT THE FOREGOING ENTRIES ARE TRUE AND CORRECT:

BY: _____ DATE: _____

DRONE OPERATION DRAWING SHEETS

DRONE OPERATION DRAWING

DATE _____

Use the area below to visually depict the flying field, flight path, flight logistics, obstacles and other details related to the flight.

FLIGHT NO. _____

COMMENTS

REFERENCE KEY
CR   Calibration Required
HZ   Hazard
PP   People and Property
AT   Air Traffic
MI   Mechanical Issue
TI   Technical Issue
PIC   Pilot in Command

DRONE OPERATION DRAWING

DATE _____

Use the area below to visually depict the flying field, flight path, flight logistics, obstacles and other details related to the flight.

FLIGHT NO. _____

COMMENTS

REFERENCE KEY
CR   Calibration Required
HZ   Hazard
PP   People and Property
AT   Air Traffic
MI   Mechanical Issue
TI   Technical Issue
PIC   Pilot in Command

DRONE OPERATION DRAWING

Use the area below to visually depict the flying field, flight path, flight logistics, obstacles and other details related to the flight.

DATE _____

FLIGHT NO. _____

COMMENTS

REFERENCE KEY
- CR — Calibration Required
- HZ — Hazard
- PP — People and Property
- AT — Air Traffic
- MI — Mechanical Issue
- TI — Technical Issue
- PIC — Pilot in Command

DRONE OPERATION DRAWING

Use the area below to visually depict the flying field, flight path, flight logistics, obstacles and other details related to the flight.

DATE _____

FLIGHT NO. _____

COMMENTS

REFERENCE KEY
CR   Calibration Required
HZ   Hazard
PP   People and Property
AT   Air Traffic
MI   Mechanical Issue
TI   Technical Issue
PIC   Pilot in Command

DRONE OPERATION DRAWING

Use the area below to visually depict the flying field, flight path, flight logistics, obstacles and other details related to the flight.

DATE _____

FLIGHT NO. _____

COMMENTS

REFERENCE KEY
CR Calibration Required
HZ Hazard
PP People and Property
AT Air Traffic
MI Mechanical Issue
TI Technical Issue
PIC Pilot in Command

DRONE OPERATION DRAWING

DATE _____

Use the area below to visually depict the flying field, flight path, flight logistics, obstacles and other details related to the flight.

FLIGHT NO. _____

COMMENTS

REFERENCE KEY
CR — Calibration Required
HZ — Hazard
PP — People and Property
AT — Air Traffic
MI — Mechanical Issue
TI — Technical Issue
PIC — Pilot in Command

DRONE OPERATION DRAWING

DATE _____

Use the area below to visually depict the flying field, flight path, flight logistics, obstacles and other details related to the flight.

FLIGHT NO. _____

COMMENTS

REFERENCE KEY
CR   Calibration Required
HZ   Hazard
PP   People and Property
AT   Air Traffic
MI   Mechanical Issue
TI   Technical Issue
PIC   Pilot in Command

DRONE OPERATION DRAWING

DATE _____

FLIGHT NO. _____

Use the area below to visually depict the flying field, flight path, flight logistics, obstacles and other details related to the flight.

COMMENTS

REFERENCE KEY
CR   Calibration Required
HZ   Hazard
PP   People and Property
AT   Air Traffic
MI   Mechanical Issue
TI   Technical Issue
PIC Pilot in Command

DRONE OPERATION DRAWING

DATE _____

Use the area below to visually depict the flying field, flight path, flight logistics, obstacles and other details related to the flight.

FLIGHT NO. _____

COMMENTS

REFERENCE KEY
CR   Calibration Required
HZ   Hazard
PP   People and Property
AT   Air Traffic
MI   Mechanical Issue
TI   Technical Issue
PIC Pilot in Command

DRONE OPERATION DRAWING

DATE _____

Use the area below to visually depict the flying field, flight path, flight logistics, obstacles and other details related to the flight.

FLIGHT NO. _____

COMMENTS

REFERENCE KEY
CR   Calibration Required
HZ   Hazard
PP   People and Property
AT   Air Traffic
MI   Mechanical Issue
TI   Technical Issue
PIC   Pilot in Command

DRONE OPERATION DRAWING

DATE _____

Use the area below to visually depict the flying field, flight path, flight logistics, obstacles and other details related to the flight.

FLIGHT NO. _____

COMMENTS

REFERENCE KEY
CR   Calibration Required
HZ   Hazard
PP   People and Property
AT   Air Traffic
MI   Mechanical Issue
TI   Technical Issue
PIC   Pilot in Command

DRONE OPERATION DRAWING

DATE _____

Use the area below to visually depict the flying field, flight path, flight logistics, obstacles and other details related to the flight.

FLIGHT NO. _____

COMMENTS

REFERENCE KEY
CR   Calibration Required
HZ   Hazard
PP   People and Property
AT   Air Traffic
MI   Mechanical Issue
TI   Technical Issue
PIC   Pilot in Command

DRONE OPERATION DRAWING

DATE _____

Use the area below to visually depict the flying field, flight path, flight logistics, obstacles and other details related to the flight.

FLIGHT NO. _____

COMMENTS

REFERENCE KEY
CR   Calibration Required
HZ   Hazard
PP   People and Property
AT   Air Traffic
MI   Mechanical Issue
TI   Technical Issue
PIC Pilot in Command

DRONE OPERATION DRAWING

DATE _____

FLIGHT NO. _____

Use the area below to visually depict the flying field, flight path, flight logistics, obstacles and other details related to the flight.

COMMENTS

REFERENCE KEY
CR   Calibration Required
HZ   Hazard
PP   People and Property
AT   Air Traffic
MI   Mechanical Issue
TI   Technical Issue
PIC Pilot in Command

DRONE OPERATION DRAWING

DATE _____

Use the area below to visually depict the flying field, flight path, flight logistics, obstacles and other details related to the flight.

FLIGHT NO. _____

COMMENTS

REFERENCE KEY
CR   Calibration Required
HZ   Hazard
PP   People and Property
AT   Air Traffic
MI   Mechanical Issue
TI   Technical Issue
PIC   Pilot in Command

DRONE OPERATION DRAWING

DATE _____

Use the area below to visually depict the flying field, flight path, flight logistics, obstacles and other details related to the flight.

FLIGHT NO. _____

COMMENTS

REFERENCE KEY
CR Calibration Required
HZ Hazard
PP People and Property
AT Air Traffic
MI Mechanical Issue
TI Technical Issue
PIC Pilot in Command

DRONE OPERATION DRAWING

DATE _____

FLIGHT NO. _____

Use the area below to visually depict the flying field, flight path, flight logistics, obstacles and other details related to the flight.

COMMENTS

REFERENCE KEY
CR   Calibration Required
HZ   Hazard
PP   People and Property
AT   Air Traffic
MI   Mechanical Issue
TI   Technical Issue
PIC   Pilot in Command

DRONE OPERATION DRAWING

DATE _____

FLIGHT NO. _____

Use the area below to visually depict the flying field, flight path, flight logistics, obstacles and other details related to the flight.

COMMENTS

REFERENCE KEY
CR   Calibration Required
HZ   Hazard
PP   People and Property
AT   Air Traffic
MI   Mechanical Issue
TI   Technical Issue
PIC Pilot in Command

DRONE OPERATION DRAWING

DATE _____

Use the area below to visually depict the flying field, flight path, flight logistics, obstacles and other details related to the flight.

FLIGHT NO. _____

COMMENTS

REFERENCE KEY
- CR — Calibration Required
- HZ — Hazard
- PP — People and Property
- AT — Air Traffic
- MI — Mechanical Issue
- TI — Technical Issue
- PIC — Pilot in Command

DRONE OPERATION DRAWING

DATE _____

FLIGHT NO. _____

Use the area below to visually depict the flying field, flight path, flight logistics, obstacles and other details related to the flight.

COMMENTS

REFERENCE KEY
CR   Calibration Required
HZ   Hazard
PP   People and Property
AT   Air Traffic
MI   Mechanical Issue
TI   Technical Issue
PIC   Pilot in Command

DRONE OPERATION DRAWING

DATE _____

FLIGHT NO. _____

Use the area below to visually depict the flying field, flight path, flight logistics, obstacles and other details related to the flight.

COMMENTS

REFERENCE KEY
CR   Calibration Required
HZ   Hazard
PP   People and Property
AT   Air Traffic
MI   Mechanical Issue
TI   Technical Issue
PIC   Pilot in Command

DRONE OPERATION DRAWING

DATE _____

FLIGHT NO. _____

Use the area below to visually depict the flying field, flight path, flight logistics, obstacles and other details related to the flight.

COMMENTS

REFERENCE KEY
CR   Calibration Required
HZ   Hazard
PP   People and Property
AT   Air Traffic
MI   Mechanical Issue
TI   Technical Issue
PIC Pilot in Command

DRONE OPERATION DRAWING

DATE _____

Use the area below to visually depict the flying field, flight path, flight logistics, obstacles and other details related to the flight.

FLIGHT NO. _____

COMMENTS

REFERENCE KEY
CR   Calibration Required
HZ   Hazard
PP   People and Property
AT   Air Traffic
MI   Mechanical Issue
TI   Technical Issue
PIC   Pilot in Command

DRONE OPERATION DRAWING

DATE _____

Use the area below to visually depict the flying field, flight path, flight logistics, obstacles and other details related to the flight.

FLIGHT NO. _____

COMMENTS

REFERENCE KEY
CR   Calibration Required
HZ   Hazard
PP   People and Property
AT   Air Traffic
MI   Mechanical Issue
TI   Technical Issue
PIC   Pilot in Command

DRONE OPERATION DRAWING

DATE _____

Use the area below to visually depict the flying field, flight path, flight logistics, obstacles and other details related to the flight.

FLIGHT NO. _____

COMMENTS

REFERENCE KEY
CR — Calibration Required
HZ — Hazard
PP — People and Property
AT — Air Traffic
MI — Mechanical Issue
TI — Technical Issue
PIC — Pilot in Command

DRONE OPERATION DRAWING

DATE _____

Use the area below to visually depict the flying field, flight path, flight logistics, obstacles and other details related to the flight.

FLIGHT NO. _____

COMMENTS

REFERENCE KEY
CR   Calibration Required
HZ   Hazard
PP   People and Property
AT   Air Traffic
MI   Mechanical Issue
TI   Technical Issue
PIC   Pilot in Command

DRONE OPERATION DRAWING

DATE _____

FLIGHT NO. _____

Use the area below to visually depict the flying field, flight path, flight logistics, obstacles and other details related to the flight.

COMMENTS

REFERENCE KEY
CR   Calibration Required
HZ   Hazard
PP   People and Property
AT   Air Traffic
MI   Mechanical Issue
TI   Technical Issue
PIC Pilot in Command

DRONE OPERATION DRAWING

DATE _____

FLIGHT NO. _____

Use the area below to visually depict the flying field, flight path, flight logistics, obstacles and other details related to the flight.

COMMENTS

REFERENCE KEY
CR	Calibration Required
HZ	Hazard
PP	People and Property
AT	Air Traffic
MI	Mechanical Issue
TI	Technical Issue
PIC	Pilot in Command

DRONE OPERATION DRAWING

DATE _____

FLIGHT NO. _____

Use the area below to visually depict the flying field, flight path, flight logistics, obstacles and other details related to the flight.

COMMENTS

REFERENCE KEY
CR   Calibration Required
HZ   Hazard
PP   People and Property
AT   Air Traffic
MI   Mechanical Issue
TI   Technical Issue
PIC   Pilot in Command

DRONE OPERATION DRAWING

DATE _____

Use the area below to visually depict the flying field, flight path, flight logistics, obstacles and other details related to the flight.

FLIGHT NO. _____

COMMENTS

REFERENCE KEY

CR	Calibration Required
HZ	Hazard
PP	People and Property
AT	Air Traffic
MI	Mechanical Issue
TI	Technical Issue
PIC	Pilot in Command

DRONE OPERATION DRAWING

DATE _____

Use the area below to visually depict the flying field, flight path, flight logistics, obstacles and other details related to the flight.

FLIGHT NO. _____

COMMENTS

REFERENCE KEY
CR   Calibration Required
HZ   Hazard
PP   People and Property
AT   Air Traffic
MI   Mechanical Issue
TI   Technical Issue
PIC   Pilot in Command

DRONE OPERATION DRAWING

DATE _____

FLIGHT NO. _____

Use the area below to visually depict the flying field, flight path, flight logistics, obstacles and other details related to the flight.

COMMENTS

REFERENCE KEY
CR   Calibration Required
HZ   Hazard
PP   People and Property
AT   Air Traffic
MI   Mechanical Issue
TI   Technical Issue
PIC Pilot in Command

DRONE OPERATION DRAWING

DATE _____

Use the area below to visually depict the flying field, flight path, flight logistics, obstacles and other details related to the flight.

FLIGHT NO. _____

COMMENTS

REFERENCE KEY
CR   Calibration Required
HZ   Hazard
PP   People and Property
AT   Air Traffic
MI   Mechanical Issue
TI   Technical Issue
PIC   Pilot in Command

DRONE OPERATION DRAWING

DATE _____

FLIGHT NO. _____

Use the area below to visually depict the flying field, flight path, flight logistics, obstacles and other details related to the flight.

COMMENTS

REFERENCE KEY
CR   Calibration Required
HZ   Hazard
PP   People and Property
AT   Air Traffic
MI   Mechanical Issue
TI   Technical Issue
PIC   Pilot in Command

DRONE OPERATION DRAWING

DATE _____

FLIGHT NO. _____

Use the area below to visually depict the flying field, flight path, flight logistics, obstacles and other details related to the flight.

COMMENTS

REFERENCE KEY
CR   Calibration Required
HZ   Hazard
PP   People and Property
AT   Air Traffic
MI   Mechanical Issue
TI   Technical Issue
PIC   Pilot in Command

DRONE OPERATION DRAWING

DATE _____

Use the area below to visually depict the flying field, flight path, flight logistics, obstacles and other details related to the flight.

FLIGHT NO. _____

COMMENTS

REFERENCE KEY
- CR Calibration Required
- HZ Hazard
- PP People and Property
- AT Air Traffic
- MI Mechanical Issue
- TI Technical Issue
- PIC Pilot in Command

DRONE OPERATION DRAWING

DATE _____

Use the area below to visually depict the flying field, flight path, flight logistics, obstacles and other details related to the flight.

FLIGHT NO. _____

COMMENTS

REFERENCE KEY
CR   Calibration Required
HZ   Hazard
PP   People and Property
AT   Air Traffic
MI   Mechanical Issue
TI   Technical Issue
PIC Pilot in Command

DRONE OPERATION DRAWING

DATE _____

FLIGHT NO. _____

Use the area below to visually depict the flying field, flight path, flight logistics, obstacles and other details related to the flight.

COMMENTS

REFERENCE KEY
CR   Calibration Required
HZ   Hazard
PP   People and Property
AT   Air Traffic
MI   Mechanical Issue
TI   Technical Issue
PIC   Pilot in Command

DRONE OPERATION DRAWING

DATE _____

Use the area below to visually depict the flying field, flight path, flight logistics, obstacles and other details related to the flight.

FLIGHT NO. _____

COMMENTS

REFERENCE KEY
- CR   Calibration Required
- HZ   Hazard
- PP   People and Property
- AT   Air Traffic
- MI   Mechanical Issue
- TI   Technical Issue
- PIC   Pilot in Command

DRONE OPERATION DRAWING

DATE _____

Use the area below to visually depict the flying field, flight path, flight logistics, obstacles and other details related to the flight.

FLIGHT NO. _____

COMMENTS

REFERENCE KEY
CR Calibration Required
HZ Hazard
PP People and Property
AT Air Traffic
MI Mechanical Issue
TI Technical Issue
PIC Pilot in Command

DRONE OPERATION DRAWING

DATE _____

Use the area below to visually depict the flying field, flight path, flight logistics, obstacles and other details related to the flight.

FLIGHT NO. _____

COMMENTS

REFERENCE KEY
- CR — Calibration Required
- HZ — Hazard
- PP — People and Property
- AT — Air Traffic
- MI — Mechanical Issue
- TI — Technical Issue
- PIC — Pilot in Command

DRONE OPERATION DRAWING

DATE _____

Use the area below to visually depict the flying field, flight path, flight logistics, obstacles and other details related to the flight.

FLIGHT NO. _____

COMMENTS

REFERENCE KEY
CR   Calibration Required
HZ   Hazard
PP   People and Property
AT   Air Traffic
MI   Mechanical Issue
TI   Technical Issue
PIC Pilot in Command

DRONE OPERATION DRAWING

DATE _____

Use the area below to visually depict the flying field, flight path, flight logistics, obstacles and other details related to the flight.

FLIGHT NO. _____

COMMENTS

REFERENCE KEY
CR   Calibration Required
HZ   Hazard
PP   People and Property
AT   Air Traffic
MI   Mechanical Issue
TI   Technical Issue
PIC   Pilot in Command

DRONE OPERATION DRAWING

Use the area below to visually depict the flying field, flight path, flight logistics, obstacles and other details related to the flight.

DATE _____

FLIGHT NO. _____

COMMENTS

REFERENCE KEY
CR   Calibration Required
HZ   Hazard
PP   People and Property
AT   Air Traffic
MI   Mechanical Issue
TI   Technical Issue
PIC   Pilot in Command

DRONE OPERATION DRAWING

DATE _____

FLIGHT NO. _____

Use the area below to visually depict the flying field, flight path, flight logistics, obstacles and other details related to the flight.

COMMENTS

REFERENCE KEY
CR	Calibration Required
HZ	Hazard
PP	People and Property
AT	Air Traffic
MI	Mechanical Issue
TI	Technical Issue
PIC	Pilot in Command

DRONE OPERATION DRAWING

DATE _____

Use the area below to visually depict the flying field, flight path, flight logistics, obstacles and other details related to the flight.

FLIGHT NO. _____

COMMENTS

REFERENCE KEY
CR   Calibration Required
HZ   Hazard
PP   People and Property
AT   Air Traffic
MI   Mechanical Issue
TI   Technical Issue
PIC Pilot in Command

DRONE OPERATION DRAWING

Use the area below to visually depict the flying field, flight path, flight logistics, obstacles and other details related to the flight.

DATE _____

FLIGHT NO. _____

COMMENTS

REFERENCE KEY
CR — Calibration Required
HZ — Hazard
PP — People and Property
AT — Air Traffic
MI — Mechanical Issue
TI — Technical Issue
PIC — Pilot in Command

DRONE OPERATION DRAWING

DATE _____

FLIGHT NO. _____

Use the area below to visually depict the flying field, flight path, flight logistics, obstacles and other details related to the flight.

COMMENTS

REFERENCE KEY
CR   Calibration Required
HZ   Hazard
PP   People and Property
AT   Air Traffic
MI   Mechanical Issue
TI   Technical Issue
PIC Pilot in Command

DRONE OPERATION DRAWING

DATE _____

Use the area below to visually depict the flying field, flight path, flight logistics, obstacles and other details related to the flight.

FLIGHT NO. _____

COMMENTS

REFERENCE KEY
CR   Calibration Required
HZ   Hazard
PP   People and Property
AT   Air Traffic
MI   Mechanical Issue
TI   Technical Issue
PIC   Pilot in Command

www.ingramcontent.com/pod-product-compliance
Lightning Source LLC
Chambersburg PA
CBHW081730170526
45167CB00009B/3766